CONTENTS

I AM A CYGNET.

I am a cygnet (SIG-nuht).
I am a baby swan.

baby SWANS

MEG GREVE

CREATIVE EDUCATION • CREATIVE PAPERBACKS

CONT

See my feathers? They are fluffy and gray.

I was in an egg for 35 days.
Then, I hatched!

My mom keeps us safe in the <u>nest</u>. We eat <u>snails</u> and bugs.

I leave the nest to go swimming.
I stay close to my family.

My mom's feet
stir the water.
This helps
me find food.

I am four
months old.
I learn to fly.

In a year, I will grow long white feathers.

I will fly away to find my own home.

SPEAK AND LISTEN
cHIR

Can you speak like a cygnet?

Baby swans chirp and peep.

Listen to these sounds:

https://www.youtube.com/
watch?v=nR6qBs03zWw

CYGNET WORDS

feathers: light, soft parts that cover a bird's body and help it fly

hatched: broke out of an egg to be born

nest: a home built by birds to keep their eggs and babies safe

snails: small animals with slimy bodies and spiral shells on their backs

READING CORNER

Andrus, Aubre. *In the Pond*. Washington, D.C.: National Geographic Kids, 2022.

Bodden, Valerie. *Swans (Amazing Animals)*. Mankato, Minn.: Creative Paperbacks, 2022.

Dumrauf, Doris. *Life in a Wetland*. Coraopolis, Penn.: Raccoon Creek Press, 2020.

INDEX

**PUBLISHED BY CREATIVE EDUCATION
AND CREATIVE PAPERBACKS**
P.O. Box 227, Mankato, Minnesota 56002
Creative Education and Creative Paperbacks
are imprints of The Creative Company
www.thecreativecompany.us

**LIBRARY OF CONGRESS CATALOGING-
IN-PUBLICATION DATA**
Names: Greve, Meg, author.
Title: Baby swans / Meg Greve.
Description: Mankato, Minnesota : Creative Education
  and Creative Paperbacks, [2025] | Series: Starting out
  | Includes bibliographical references and index. |
  Audience: Ages 4-7 | Audience: Grades K-1 | Summary:
  "Introduce beginning readers to the webbed world of
  baby swans with this life science starter. Includes bird
  photos, a labeled animal diagram, "Make a Noise"
  section, glossary, and further resources"-- Provided by
  publisher.
Identifiers: LCCN 2024014686 (print) | LCCN 2024014687
  (ebook) | ISBN 9781640264250 (library binding) | ISBN
  9781628329582 (paperback) | ISBN 9781640005891
  (ebook)
Subjects: LCSH: Swans--Juvenile literature. | Swans--
  Infancy--Juvenile literature.
Classification: LCC QL696.A52 G738 2025 (print) |
  LCC QL696.A52 (ebook) | DDC 598.4/181392--dc23/
  eng/20240419
LC record available at https://lccn.loc.gov/2024014686
LC ebook record available at https://lccn.loc.
  gov/2024014687

DESIGN AND PRODUCTION
Design by Rhea Magaro
Production by Beeline Media and Design, Inc.
Art direction by Tom Morgan

PHOTOGRAPHS by Alamy Stock Photo/blickwinkel/
McPHOTO/PUM, 10, John Davidson Photos, 8, Krys Bailey, 6,
Tim Mason, 11; Getty Images/tracielouise, 5; Shutterstock/
Andreas Nesslinger, 7, Armensl, 4, Bohbeh, 13, Chutharat
Kamkhuntee, 2, Eric Isselee, cover, 12, Janice Chen, 9,
Tanya_Terekhina, 1

Printed in China